Introduction

Many different types of materials exist. They have different properties and are used for many different jobs.

Making new materials is important because new materials have new properties. They may be more useful than existing materials.

▼ All materials are made from natural materials. Natural materials are found in the earth, in the air, in the sea and in living things.

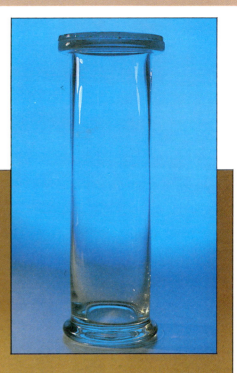

Q1 List all the natural materials shown on this page.

Q2 Say where each natural material is found.

▼ All these materials are made from the natural materials shown on page 1.

Q3 Why is it important to make new materials?

Q4 List all the synthetic materials, shown above, which come from oil.

Q5 List all the synthetic materials, shown above, which come from limestone.

Extension exercises 1–3 can be used now.

1 Chemical reactions

Chemical reactions are permanent changes

In **chemical reactions** materials are changed permanently into new materials. The new material will have new properties.

▲ The rusting of iron is a permanent change. It is a slow chemical reaction.

▲ Cooking often involves permanent changes. These changes are chemical reactions.

▼ Food going bad involves slow chemical reactions.

▲ An exploding bomb involves a very fast chemical reaction.

Q1 What does the phrase *chemical reactions* mean?

Q2 Give two examples of chemical reactions.

What is a chemical reaction?

Chemical reactions make new materials.

In this experiment you will observe what happens to materials when they are heated. Your observations will help you to understand more about what is meant by a *chemical reaction*.

Q1 Copy this table.

Table 1

Name of material	Appearance		
	Before	During	After

Apparatus

- ☐ Bunsen burner
- ☐ tripod and gauze ☐ tongs
- ☐ eye protection
- ☐ selection of materials
- ☐ small test tubes
- ☐ test tube holders
- ☐ heatproof mat ☐ spatula

When heating materials:
Wear eye protection
Point the test tube away from everybody
Let things cool on a heatproof mat.

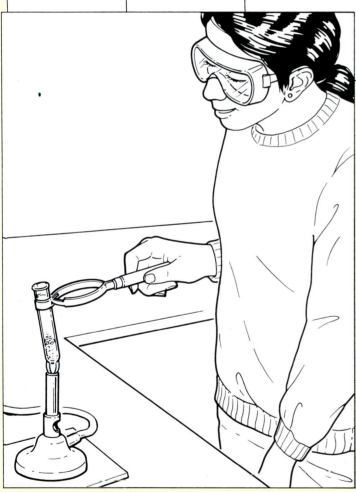

A If the material is a powder, put a spatula full into a small test tube. Heat in a medium Bunsen flame. ▲

B If the material is a solid, use tongs to hold it in a medium Bunsen flame. ▲

C Choose a material. Complete columns 1 and 2 of the table.

D Heat it as shown in **A** or **B**. Complete column 3.

E Leave it to cool on a mat. Complete column 4.

F Repeat **C** to **E** for each material.

Look at your results in table 1. Some materials did not change when heated. Some changed during heating but changed back on cooling. We call this a **temporary change**.

Other materials changed during heating and did not change back on cooling. We call this a **permanent change**. A chemical reaction has taken place.

Q2 Copy this table.

Table 2

Name of material	No change	Permanent change	Temporary change

▼ Are these chemical reactions?

G Complete column 1 using the materials from table 1.

H Look at table 1. If the material did not change at all, tick the *no change* column in table 2.

I If it changed during heating and changed back on cooling, tick *temporary change*.

J If it changed and stayed changed, tick *permanent change*.

Q3 What kinds of changes are chemical reactions?

Q4 Look at table 2. Make a list of all materials that changed in chemical reactions.

Q5 Say whether these changes are chemical reactions or not.

1 Boiling an egg
2 Heating water
3 Frying chips
4 Apples going brown
5 Dissolving sugar in a cup of tea
6 Bleaching a pair of jeans
7 Burning a candle
8 Jelly setting.

Extension exercise 4 can be used now.

How fast is a chemical reaction?

Chemical reactions are permanent changes.
Chemical reactions take place at different speeds.

It may be useful to change the speed of a chemical reaction. In industry a faster chemical reaction may save money.

These experiments will show you how we can change the speed of a chemical reaction.

Apparatus

☐ marble chips ☐ dilute acid
☐ 3 100 cm³ beakers
☐ 10 cm³ measuring cylinder
☐ cloth ☐ hammer
☐ Bunsen burner
☐ tripod and gauze
☐ heatproof mat
☐ thermometer
☐ eye protection

 Wear eye protection when using acids and when breaking marble.

Experiment 1: Size

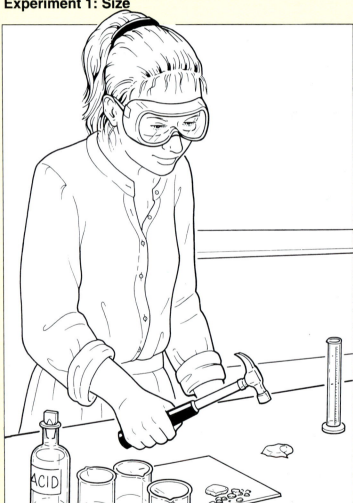

A Collect three marble chips of the same size. Break one into small pieces. Leave one whole. Use the cloth and hammer to crush the other into fine powder. ▲

B Label three beakers 1, 2, and 3. Put 10 cm³ of dilute acid into each beaker. Quickly add the whole marble chip to beaker 1, the small pieces of marble chips to beaker 2 and the powder to beaker 3. ▲

Q1 In which beaker does the marble disappear the fastest?

Q2 In which beaker is the chemical reaction the fastest?

Experiment 2: Temperature

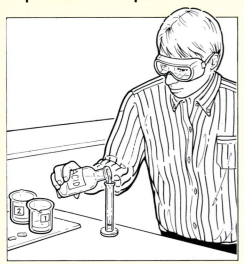

A Collect two marble chips of similar size. Label two beakers 1 and 2 and put 10 cm³ of dilute acid into each one. ▲

B Warm the acid in beaker 2 to about 40°C using a medium Bunsen flame. Turn off the Bunsen. Quickly add a marble chip to each beaker. ▲

Q3 In which beaker did the marble chip disappear first?

Q4 In which beaker was the chemical reaction the faster?

Experiment 3: Amount of acid

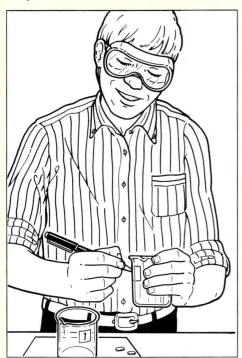

A Collect two marble chips of similar size. Label the beakers 1 and 2. ▲

B Into beaker 1, put 10 cm³ of dilute acid. Into beaker 2 put 5 cm³ of dilute acid and 5 cm³ of water. Quickly put a marble chip into each beaker and watch what happens. ▲

Q5 In which beaker is the acid more **concentrated**?

Q6 In which beaker does the marble chip disappear the fastest?

Q7 In which beaker is the chemical reaction the fastest?

Q8 Copy the sentence below and fill in the missing words.

We can make a chemical reaction go faster by making the pieces .... , by .... the temperature or by making the acid more ....

Q9 How could you make a chemical reaction go more slowly?

Q10 In each experiment you chose marble chips which were similar in size. Why was it important to choose similar marble chips?

Extension exercise 5 can be used now.

Speeding up reactions with catalysts

Catalysts are materials which speed up chemical reactions without being used up themselves. Catalysts are used in industry to save money. They allow new materials to be made faster and therefore more cheaply.

Catalysts called **enzymes** are found in living things.

Hydrogen peroxide changes to water and oxygen very slowly.

Hydrogen peroxide $\longrightarrow$ water + oxygen

Manganese dioxide is a catalyst (but not an enzyme) that speeds up this reaction. Both liver and potato contain an enzyme which speeds up the reaction.

Q1 Copy this table.

Tube number	Did the glowing spill relight?	Was oxygen made?

Apparatus

- ☐ 20 vol hydrogen peroxide solution ☐ potato slice
- ☐ piece of fresh liver
- ☐ manganese dioxide (harmful)
- ☐ large test tubes
- ☐ wood spill ☐ spatula
- ☐ gloves ☐ eye protection
- ☐ labels or chinagraph pencil
- ☐ tongs ☐ test tube rack
- ☐ Bunsen burner
- ☐ heatproof mat

 Wear eye protection when handling materials.

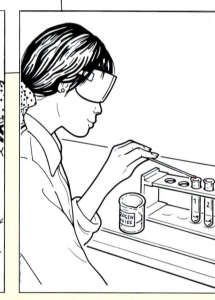

A Label four tubes 1–4. Add 1 cm depth of hydrogen peroxide to each tube. ▲

B Test tube 1 for oxygen with a glowing spill. (It relights if oxygen is present.) ▲

C Add a spatula of manganese dioxide to tube 2. Test for oxygen again. ▲

D Repeat **C–D** with a piece of liver, then with potato. ▶

Q2 Which catalyst worked best for you?

Q3 Name three catalysts that speed up the reaction of hydrogen peroxide.

Q4 What is an enzyme?

Q5 Why are catalysts important?

Looking closely at a chemical reaction: combustion

▶ When a fuel such as petrol or natural gas burns, it reacts with the oxygen in the air. We call this reaction **combustion**. Combustion is one of the most important chemical reactions. It is used to heat our homes, to make electricity and to power most forms of transport.

Unfortunately combustion has important effects on the environment. It adds to the **greenhouse effect.** Burning some fuels causes **acid rain**. Combustion uses up precious fuels.

In this experiment your teacher will show you what is made when a fuel is burnt.

▶ Limewater turns *milky* when carbon dioxide gas touches it.
▶ ▶ Dry copper sulphate is white but it turns blue when water touches it.

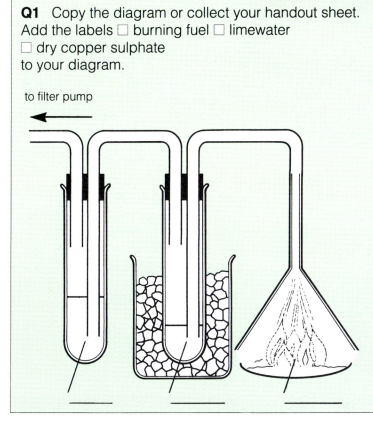

Q1 Copy the diagram or collect your handout sheet. Add the labels ☐ burning fuel ☐ limewater ☐ dry copper sulphate to your diagram.

to filter pump

Q2 Copy and complete the sentences below. The gases from the burning fuel are drawn through the apparatus. The limewater turns from clear to .... .This shows that .... is made. The dry copper sulphate turns from .... to .... This shows that .... is made.

Q3 Complete the word equation which shows what happens when a fuel is burned.
fuel + oxygen ⟶ .... + ...

Q4 The most important thing that combustion produces is not a material. It is not shown in the equation. What is it?

Q5 The carbon dioxide produced in combustion contributes to the greenhouse effect.
Find out about the greenhouse effect and write a short report about it.

Q6 Acid rain is also made when some fuels burn. Find out about acid rain and write a short report about it.

Do all chemical reactions give out heat?

Combustion is a chemical reaction that gives out heat energy. In this experiment we shall see if this is true of all chemical reactions.

We shall use a thermometer. If the reaction gives out heat the temperature will rise. If it takes in heat the temperature will fall.

Q1 Copy this table.

Reaction	Temperature before	Temperature after	Heat taken in or given out

A Your teacher will show you some chemical reactions. Complete your table.

B Put 25 cm³ of copper sulphate solution into a polystyrene cup. Take its temperature. Add a spatula full of iron filings. Stir, and take the temperature. Complete the table. ▲

C Put 25 cm³ of dilute acid into a clean polystyrene cup. Take the temperature. Add the piece of magnesium ribbon. When the reaction has finished, take the temperature again. Complete the table. ▲

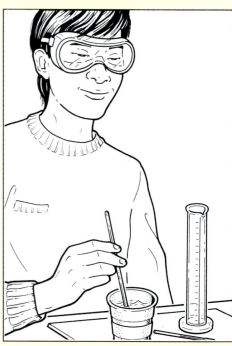

D Put 25 cm³ of water into a polystyrene cup. Take its temperature. Add the ammonium sulphate. Stir. When no more white powder can be seen, take the temperature again. Complete the table. ▲

Q2 Copy out the sentence which best describes the results of your experiments
a all chemical reactions give out heat
b most chemical reactions give out heat
c only a few chemical reactions give out heat.

Extension exercise 6 can be used now.

Respiration

On page 9 you learnt that oxygen is used up when a fuel is burnt (combustion). Carbon dioxide and water are the new materials made. Heat energy is also produced.

Combustion   fuel + oxygen ⟶ carbon dioxide + water + ENERGY

When living things breathe in they take in oxygen. When they breathe out, they give out carbon dioxide and water. Using up oxygen and making carbon dioxide and water is part of **respiration**.

▶ Respiration also uses up food. The reason for respiration is to release energy from food. Energy from food is needed by all living things. Respiration is a chemical reaction.

Respiration   food + oxygen ⟶ carbon dioxide + water + ENERGY

▶ Combustion and respiration take place at different temperatures. Fuels need to be heated to start them burning. Respiration takes place in living things at lower temperatures. Enzymes (see page 8) are used to help respiration work at lower temperatures.

Q1 What new materials are made in both respiration and combustion?

Q2 What is used up in both reactions?

Q3 What difference is there between respiration and combustion?

Q4 What is the most useful product of each reaction?

What turns fresh apples brown?

This experiment lets you find out what turns fresh apples brown.
Apples go brown when cut and left in the open air. This experiment
tests the idea that air causes apples to go brown.

Q1 Copy this table.

Tube number	Conditions (amount of air)	Amount of browning after 15 mins

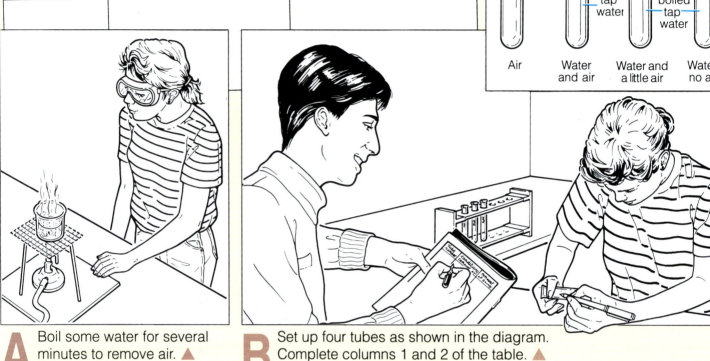

Air | Water and air | Water and a little air | Water, no air

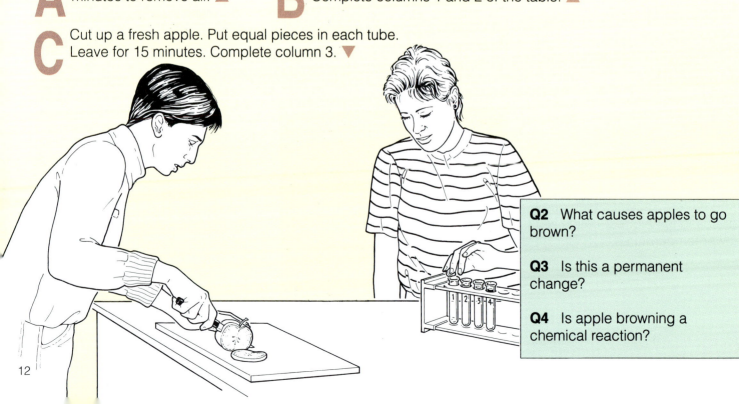

A Boil some water for several minutes to remove air. ▲

B Set up four tubes as shown in the diagram. Complete columns 1 and 2 of the table. ▲

C Cut up a fresh apple. Put equal pieces in each tube. Leave for 15 minutes. Complete column 3. ▼

Q2 What causes apples to go brown?

Q3 Is this a permanent change?

Q4 Is apple browning a chemical reaction?

Stopping apples going brown

When food goes bad chemical reactions change part of the food into harmful materials. We can add materials to food to stop or slow down the process of it going bad. The materials we use must be harmless.

In this experiment you are asked to find out which materials stop apples turning brown. You can try these materials: water, sugar solution, lemon juice, vinegar.

You must *design* the experiment yourself. Use page 12 to help you. It must be a fair test.

A Write down instructions for your experiment.

B Draw a diagram showing the experiment.

C Discuss your plan with your teacher. ▼

D Do the experiment.

E Write a report. It should say what you did and show your results.

2 The activity series of metals

Metals are one of the most useful types of material.

In this chapter we will look at some of the chemical reactions of the metals potassium, sodium, calcium, magnesium, aluminium, zinc, iron, copper, silver and gold. We will then use these reactions to put the metals in the order of how quickly they react. The list of metals in order of how quickly they react is called the **activity series**.

Knowing about the activity series helps us to understand how to get metals from the earth. It also helps us to understand why some metals **corrode** and how to prevent corrosion.

The reactions of potassium, sodium and calcium with cold water

Q1 Copy this table.

Name of metal	Does it move on water?	How does it burn?	Does the reaction make hydrogen?	Does the reaction make an alkali

A Your teacher will show you the reactions of sodium and potassium with cold water. Watch carefully and complete the table.

B Half fill a beaker with cold water. Fill the test tube with cold water and put it upside down in the beaker so that none of the water falls out. ▶

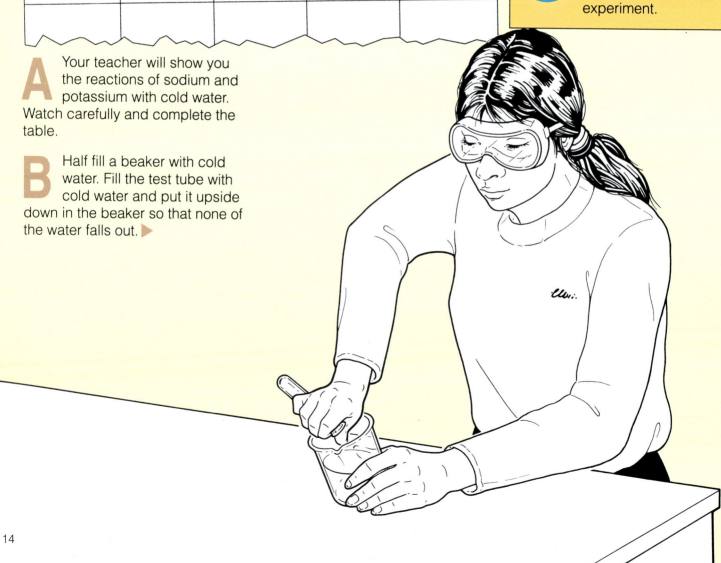

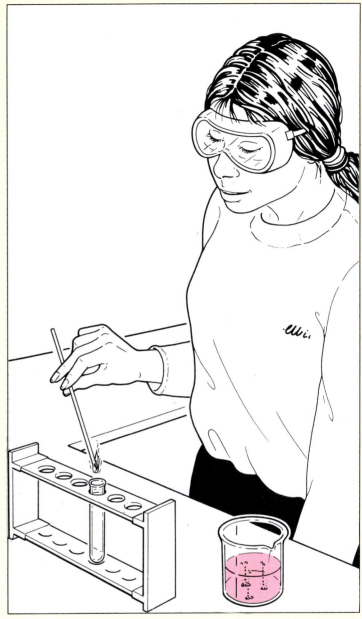

C Use tweezers to add a few pieces of calcium to the beaker. Put the test tube of water over one of the pieces of calcium. ▲

D When the test tube is full of gas, take it out of the beaker. Put a lit spill to the end of the test tube. Put a few drops of phenolphthalein indicator into the beaker of water. Complete the last column of your table. ▲

Q2 What is the name of the gas made when calcium reacts with cold water?

Q3 How do you know that a new material has been made in the water?

Q4 Put the metals sodium, potassium and calcium in order. Start with the one that reacts most quickly with cold water.

Q5 Potassium and sodium are called **alkali metals**. Why?

Q6 Copy:
Metals that react with cold water follow the same pattern.
metal + water ⟶ alkali + hydrogen

Q7 What does corrode mean?

The reaction of magnesium with cold water

We will look at the reaction of another metal, magnesium, with cold water.

Apparatus

☐ glass funnel
☐ clean and shiny magnesium ribbon (flammable)
☐ test tube ☐ 250 cm³ beaker
☐ phenolpthalein indicator (flammable)

A Fill the beaker with cold water to the 200 cm³ mark. Add a few drops of the indicator. ▼

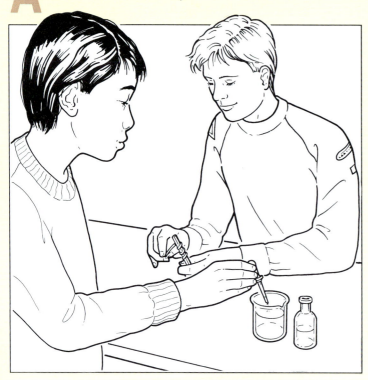

B Put the magnesium ribbon into the beaker under the funnel. Fill the test tube with cold water and put it over the funnel so that no water spills out. ▼

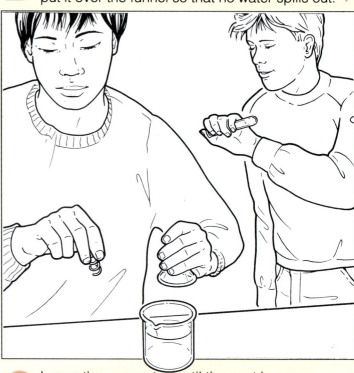

C Leave the apparatus until the next lesson.

D Look at the colour of the solution around the magnesium. Test any gas that has collected in the test tube with a lighted spill. ◄

Q1 Draw two diagrams showing the apparatus *before* and then *after* the reaction.

Q2 How do you know that a reaction has taken place?

Q3 Does the magnesium react quickly or slowly with cold water?

Q4 Put the metals magnesium, sodium, potassium and calcium in order of how quickly they react with cold water.

Extension exercise 7 can be used now.

The reactions of metals with acids

Some metals, like sodium and potassium, react quickly with cold water. Other metals, like magnesium, react slowly with cold water. Metals that react slowly may react more quickly with dilute acid. This experiment will help you put the metals in order of how quickly they react with acid.

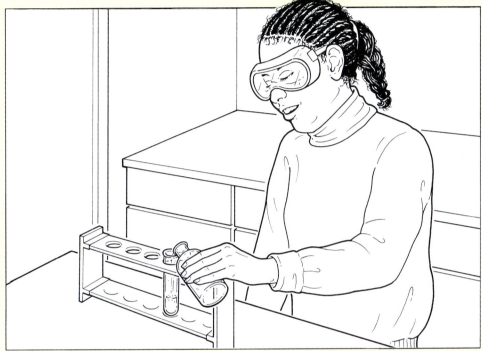

A Put 2 cm depth of dilute acid into a large test tube. ▲

B Put a piece of magnesium into the test tube. Test for hydrogen. ▼

C Try each metal in turn with acid. You may have to warm the acid to make hydrogen.

D Design your own results table and fill it in.

You should now be able to put the metals sodium, potassium, calcium, magnesium, iron, zinc and copper in order of how quickly they will react with cold water and with acid.

Check your list's order with the list printed inside the back cover.

Look at the inside back cover and answer these questions.

Q1 Which is the most reactive metal?

Q2 Which is the least reactive metal?

Q3 Which metals will not react with dilute acid?

Q4 Nickel is a common metal missing from the list. What experiments would you do to find out where it should go in the list?

The Thermit reaction

▶ Your teacher will show you (from a safe distance!) a chemical reaction called the **Thermit reaction**.

◀ The Thermit reaction in action

Q1 What did you see as the reaction took place?

Q2 Name the two materials that reacted together.

Q3 What was found at the end of the experiment?

Explaining the Thermit reaction

In the Thermit reaction, aluminium and iron oxide were mixed together. When the mixture was heated a chemical reaction took place. The aluminium gained oxygen to become aluminium oxide. The iron oxide lost oxygen to become iron.

We can use a word equation to show this reaction.

aluminium + iron oxide ⟶ aluminium oxide + iron

We say that aluminium is **oxidised** because it gains oxygen.
We say that iron oxide is **reduced** because it loses oxygen.

A chemical reaction in which something is oxidised is called an **oxidation reaction**.
A chemical reaction in which something is reduced is called a **reduction reaction**.

The Thermit reaction works because aluminium is more reactive than iron. The more reactive aluminium is able to take oxygen from the iron oxide. It reduces the iron oxide to iron. It is a **reducing agent**.

The iron oxide gives up its oxygen to the aluminium. It oxidises the aluminium. It is an **oxidising agent**. A lot of heat is given out during the Thermit reaction.

Here is another example of a reaction where a more reactive metal takes oxygen from a metal oxide.

magnesium
+
copper oxide

↓

copper
+
magnesium oxide

Q4 Which has been oxidised, the magnesium or the copper oxide?

Q5 Which is the oxidising agent?

Q6 Which is the more reactive metal, magnesium or copper?

Q7 If iron oxide was mixed and heated with zinc, would a reaction take place? Explain your answer.

Q8 If calcium oxide is heated with zinc, nothing happens. Why is this?

Extension exercise 8 can be used now.

Getting copper from malachite

Malachite is an **ore** found in some rocks. It is a **compound** of copper. Malachite contains the **elements** copper, carbon and oxygen joined together. The copper can be taken out of the malachite by heating the malachite with carbon. Taking metals out of ores is called **extraction**.

Apparatus

☐ lump of malachite
☐ powdered, pure malachite
☐ large test tube
☐ Bunsen burner
☐ heatproof mat
☐ 250 cm³ beaker
☐ carbon powder
☐ test tube holder
☐ eye protection ☐ scissors
☐ 2 spatulas ☐ sticky tape

 Wear eye protection when heating materials.

A Put five spatula measures of malachite powder into a large test tube. ▼

B Hold the test tube with the holder. Heat the powder *gently* until it turns black and stops rising. Let the test tube cool. ▼

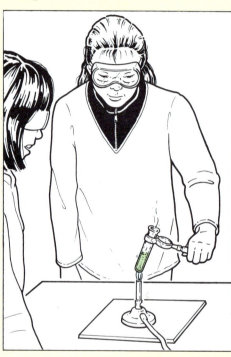

C Add two spatula measures of carbon powder and mix well. Heat the mixture *strongly* until it goes red. Turn off the Bunsen. Let the mixture cool. ▲

D Half fill the beaker with water. Pour the mixture from the test tube into the cold water. Leave it for two minutes. ▼

E Pour off the dirty water. Take care not to lose any copper. ▼

F Repeat **D** and **E** until you have pink copper at the bottom of the beaker. Put the copper in a warm place to dry. Stick it in your notebook with sticky tape.

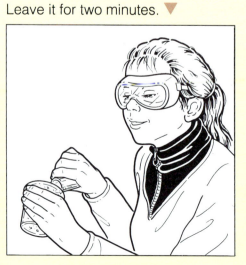

Q1 What did you see in **B** as malachite was gently heated?

Q2 What did you see in **C** as the black powders were heated strongly?

Q3 How many chemical reactions are there in **B** and **C**?

Getting metals from their ores

In the last experiment you extracted (took out) copper metal from the ore malachite. You did this by heating the pure malachite with carbon.

Malachite is a compound of copper called copper carbonate. When it is heated gently it becomes black copper oxide. The copper oxide is then heated with carbon. The carbon reduces the copper oxide to copper. We can show this with a word equation.

copper oxide + carbon $\longrightarrow$ carbon dioxide + copper

panning for gold in a river

▲ Metals at the bottom of the activity series like gold and silver are found **native**. This means they are found by themselves, not combined with other elements in ores. This is because they do not react easily with other materials.

◄ Metals at the bottom of the activity series are used to make jewellery. They are unreactive and do not corrode easily.

▲ Metals from the middle of the activity series have to be extracted from their ores. You have seen that one way of doing this is to heat the ore with carbon. Copper has been extracted in this way since earliest times.

◄ Iron is extracted in a similar way today.

Q1 Why is carbon able to reduce copper oxide? (**Hint**: Remember the activity series.)

Q2 Look at the cartoon on page 19. Draw a similar cartoon to show carbon taking oxygen from copper.

Extracting metals using electricity

Metals high up in the activity series like sodium and aluminium need a lot of energy to extract them from their ores. The ore has to be melted and then an electric current passed through it. Both the heat for melting and the electricity needed make metals like aluminium expensive.

▼ Sodium and aluminium have only been known as metals since the middle of the last century. They were only discovered when electricity could be used.

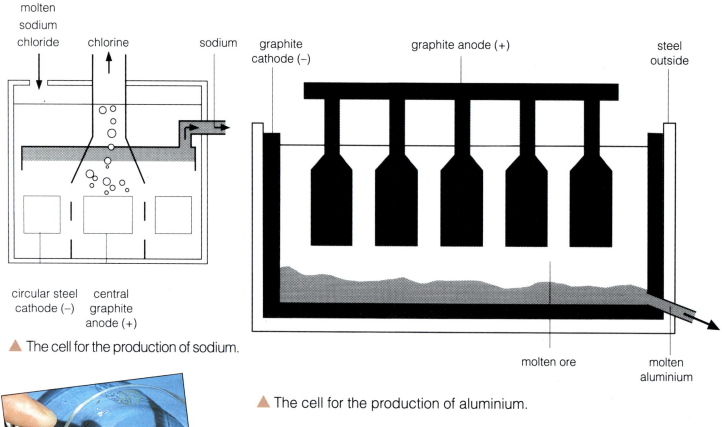

molten sodium chloride

chlorine

sodium

graphite cathode (−)

graphite anode (+)

steel outside

circular steel cathode (−)   central graphite anode (+)

▲ The cell for the production of sodium.

molten ore

molten aluminium

▲ The cell for the production of aluminium.

Q1 There is more aluminium ore in the ground than any other metal or metal ore. Why is aluminium more expensive than copper or iron?

Q2 Metals like gold and silver have been known since earliest times. Why has sodium only been known since the nineteenth century?

Q3 When do you think potassium was discovered?

Using electricity to get copper from malachite

This experiment shows you how to extract copper from malachite using electricity. The malachite is first made into copper sulphate solution by reacting it with acid.

Apparatus

☐ pure malachite (harmful)
☐ 100 cm³ beakers
☐ 25 cm³ beaker ☐ spatula
☐ filter funnel and paper
☐ conical flask ☐ two electrodes
☐ wires with clips
☐ bulb in holder ☐ dilute acid
☐ battery or power pack
☐ 25 cm³ measuring cylinder
☐ eye protection

 Eye protection must be worn when using acids.

A Put 25 cm³ of dilute acid into the 100 cm³ beaker. Add malachite until the fizzing stops. ▼

B Filter the mixture. Pour the blue liquid into a 25 cm³ beaker. ▼

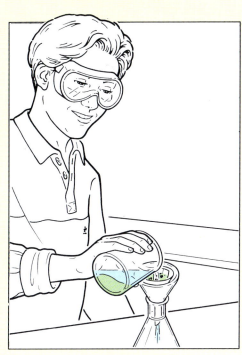

C Attach the electrodes to the battery using the wires with clips. Put the bulb between one of the electrodes and the battery. ▶

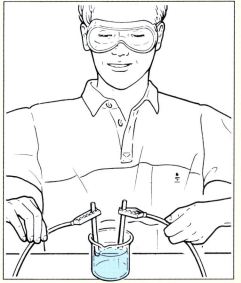

D Arrange the apparatus so that the electrodes are in the blue liquid but are not touching each other. Leave for five minutes. ▲

Q1 What is formed on the negative electrode?

Q2 Where has it come from?

Q3 What happens to the positive electrode? Do you think that this is a cheap way of getting copper from malachite?

Q4 You have seen two ways of extracting copper from malachite.
1 Heating with carbon (reduction with carbon, pages 20 – 21).
2 Using electricity (this page).
A third way was discussed on pages 18 – 19.
3 A Thermit-type reaction with magnesium.
Think about each method and try to arrange them in order of cost.

Extension exercise 9 can be used now.

The electrolysis of brine

If electricity is passed through a solution and causes a reaction we call this **electrolysis**. On page 23 you carried out the electrolysis of copper sulphate solution. It made copper metal.

In this experiment we find out what useful new materials can be made when an electric current is passed through brine (salt solution).

Q1 What did you see in the tubes as the electric current was switched on?

Q2 What colour was the gas that collected at the positive electrode (the **anode**)?

Q3 What effect did this gas have on damp red litmus paper?

Q4 What is the name of the gas?

Q5 What happened to the gas that collected at the **cathode** when a lit spill was held to the mouth of the tube?

Q6 What is the name of this gas?

Q7 What happened to the red litmus paper when it was put into the solution that was left at the end of the experiment?

Q8 Use the litmus paper to find out what material has been made.

Extension exercise 10 can be used now.

Explaining what happened

The electrolysis of brine makes three important materials. They are: *chlorine, hydrogen and sodium hydroxide solution.*

Chlorine is used to make bleach, disinfectant, plastics and solvents. It is used to kill germs in swimming baths.

Sodium hydroxide is used to make bleach, disinfectant, soap, weed killer and paper. Sodium hydroxide is a strong alkali commonly called *caustic soda*. It can be used to unblock drains.

Brine is sodium chloride dissolved in water. Sodium chloride's **chemical formula** is **NaCl**.

Water's chemical formula is H_2O or **H–OH**

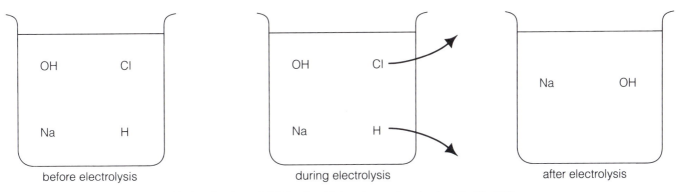

before electrolysis   during electrolysis   after electrolysis

Na (sodium) and **OH** (hydroxide) make sodium hydroxide (**NaOH**)

Making sodium hydroxide

Making salt

The chemical reaction between an acid and alkali is called **neutralisation**. When hydrochloric acid is neutralised by the alkali sodium hydroxide, sodium chloride (common salt) is made.

A Put 10 cm³ of dilute acid into a small beaker. Add three drops of universal indicator solution. ▶

Apparatus

☐ dilute hydrochloric acid
☐ dilute sodium hydroxide solution (corrosive) ☐ dropping pipette ☐ heatproof mat
☐ evaporating dish
☐ small beaker
☐ universal indicator solution
☐ 10 cm³ measuring cylinder
☐ small conical flask
☐ animal charcoal ☐ tripod
☐ filter funnel and paper
☐ Bunsen burner ☐ gauze
☐ eye protection

 Wear eye protection at all times.

 Do not taste the solid that you make.

B Add sodium hydroxide solution drop by drop to the acid in the beaker, until the colour of the solution is green. If the solution goes blue, you have added too much sodium hydroxide solution. You will need to add some more acid. ▲

C Put three spatula measures of animal charcoal into the beaker. Heat it over a medium Bunsen flame until the mixture just boils. Turn the Bunsen off and leave the beaker to cool. ▲

D When the beaker is cool, filter the mixture. Put the clear solution into the evaporating dish. Heat the dish on a water bath until all the liquid evaporates. ▶

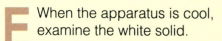

E When the apparatus is cool, examine the white solid.

Q1 What is the reaction between an acid and an alkali called?

Q2 Name the acid and name the alkali used in this experiment.

Q3 Give another name for common salt.

Q4 What did boiling the solution from **B** with animal charcoal do to the coloured indicator?

Using neutralisation

Neutralisation is a chemical reaction where an acid is used up by an alkali.

▲ Bee stings contain acid. If you are stung by a bee, put some sodium bicarbonate on it. This will help to ease the pain by neutralising the acid.
Wasp stings contain alkali. You should treat a wasp sting with a weak acid. Vinegar or lemon juice will help to neutralise the alkali.

▲ Soaps and shampoos are sometimes sold as *pH-balanced*. They are used to give the best balance of acidity/alkalinity whatever water is used for washing. Skin and hair are slightly acid. Weak acids like vinegar and lemon juice give hair an attractive shine. Hair conditioners often contain weak acids.

▲ Soil may be too acidic or alkaline for plants to grow well. Gardeners add peat to soil to make it more acidic. Lime is added to make it more alkaline (page 29).

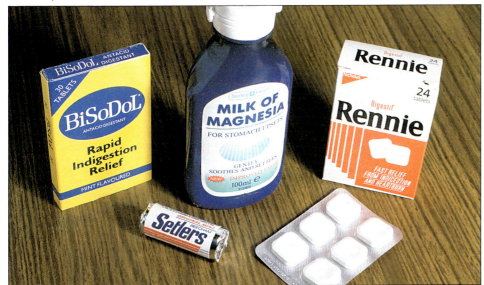

▲ Indigestion is caused by too much acid in the stomach. This can be helped by taking an **antacid** to reduce the amount of acid in the stomach. Many of these contain an alkali to neutralise the acid.

Q1 What should you put on a bee sting?

Q2 Why do some shampoos contain lemon juice?

Q3 What would you add to soil to make it more acid?

Q4 Why do many stomach powders contain alkalis?

Extension exercises 11 and 12 can be used now.

Making lime

Lime made from limestone is a white powder. It has many uses. Farmers put it on soil which is too acidic.

Limestone is calcium carbonate. Lime is calcium oxide.

Lime is made by heating limestone strongly. Strong heat breaks limestone down into two new materials.

calcium carbonate (limestone) ⟶ calcium oxide (lime) + carbon dioxide

A chemical reaction in which a material is broken down by heat is called **thermal decomposition**.

Q1 Copy this table.

	Effect of water	Effect of indicator
Limestone		
Lime		

A Place a piece of limestone onto the gauze on the corner of a tripod. Heat with a very strong flame for ten minutes to change it to lime (corrosive). Leave the apparatus to cool. ▲

B Use tongs to put the piece of lime into a test tube. Put a piece of limestone into another tube. ▲

C Carefully add a few drops of water to each tube. Notice what happens. Complete the first column in the table. ▲

D Put some universal indicator solution into each tube. Look for any colour change. Complete the second column in the table. ▶

Q2 Does limestone change the colour of the indicator solution?

Q3 Is lime an acid, an alkali or a neutral material?

Q4 Give two properties of lime which are different to those of limestone.

Q5 What is lime used for?

Limestone and lime

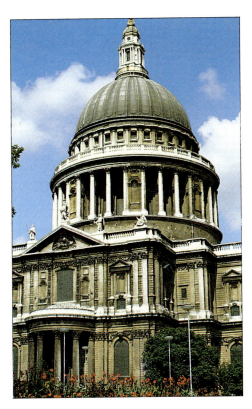

▲ Limestone is a very useful natural material. It has many important uses: making lime, making sodium carbonate, iron extraction, making cement, as a building stone.

▲ Limestone is found in many parts of Britain. The area where it is taken from the ground is called a **quarry**. Limestone quarrying has left ugly scars in certain parts of the country. Usually they are areas of great natural beauty like the Yorkshire Dales and the Peak District.

▲ Lime is made by heating limestone at very high temperatures. This is done in large kilns. A large proportion of the limestone which is quarried is turned into lime.

▲ Garden lime is used to make soil less acid. Lime neutralises acid (page 27).

Fermentation

Many chemical reactions take place in living things. Some of these can be used by people to make useful new materials. In this experiment we look at the new materials made by the chemical reactions in yeast

Yeast uses a reaction called **fermentation** to release energy from food.

Yeast can make new materials if it has food and is kept warm. Under these conditions it can grow.

Apparatus

☐ 4 test tubes
☐ fresh apple juice or sugar solution
☐ fresh yeast ☐ boiled yeast
☐ distilled water ☐ 4 balloons
☐ test tube rack
☐ chinagraph pencil or labels

Q1 Copy this table.

Tube number	Contents of tube	Appearance after 3 days	Is a gas made?	Smell

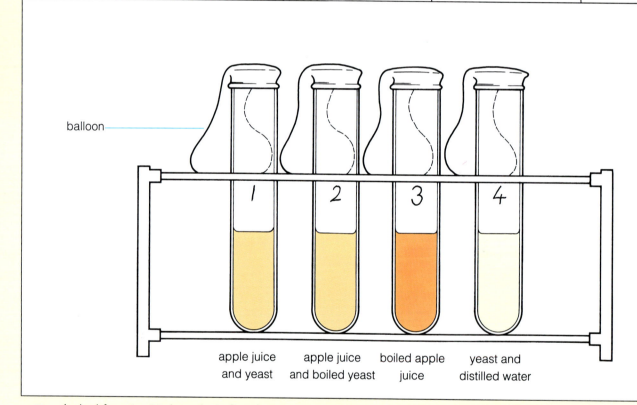

balloon

1   2   3   4

apple juice and yeast   apple juice and boiled yeast   boiled apple juice   yeast and distilled water

A Label four test tubes 1–4. Set them up as shown in the diagram, with a balloon over each tube. ▲

B Leave the test tubes several days in a warm place.

C Look at each tube. Record its appearance in the table. Complete the rest of the table.

Q2 How could you tell in which tubes a gas was made?

Q3 In which tubes could you smell a new material?

Q4 What is the name of this new material?

Q5 Why did nothing happen in tube 2?

Q6 Why were the tubes left in a warm place?

Using fermentation

Fermentation is a chemical reaction carried out by yeast to produce energy. It also makes two new materials: alcohol and carbon dioxide. They are both made from sugar.

sugar ⟶ alcohol + carbon dioxide + ENERGY

▼ Industry uses fermentation in several ways...

▲ making wine

▲ making bread rise

▲ making soy sauce

▲ making 'alcool', fuel for cars.

4 Another kind of reaction: nuclear reactions

Another kind of reaction which makes new materials is called a nuclear reaction. It is different to a chemical reaction because much more energy is given out. This energy is called radiation. Radiation can be both harmful and useful.

Some materials in the earth give off natural radiation. This is called **background radiation**. Houses in Cornwall and the east of Scotland have a higher background radiation than houses in London.

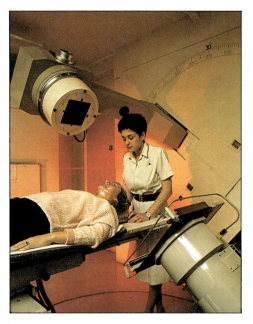

▲ Radiation is used to treat cancer.

▲ The radiation released by nuclear reactions is used to make weapons. ▶ Food can be kept fresh by using radiation. Radiation kills bacteria.

▲ Nuclear energy can be used to make electricity. Waste from nuclear power stations can pollute the environment.

Q1 Give one difference between chemical reactions and nuclear reactions.

Q2 What is background radiation?

Q3 Give one harmful effect of radiation.

Q4 Give one beneficial effect of radiation.

Q5 Why are some people against the use of nuclear energy to make electricity?